SCIENCE WITH A BEAT

OUR WONDERFUL PLANET EARTH

By Jacquie Hawkins

What is the Earth like inside? P.1

What is magma? P.4

Where does natural gas come from? P7

How is gas made in a gas plant? P.10

Why do pots on a gas stove sometimes get black? P.13

What is a desert like? P.15

Why does a geyser throw out steam? P.19

Why is there sand on a beach? P.21

Why is the water in the ocean salty? P.23

Why is there always a breeze at the ocean? P.25

What holds an island up? P.27

How were caves formed? P.29

How can wind and rain carve rocks? P.32

What are wind caves? P.34

What was the ice age? P.35

What is a glacier? P.37

What are ice caves? P.40

Why does a compass point north? P.41

CHAPTER ONE
What is the earth like inside?

The earth on the inside looks much like a baseball.
If you cut one open to look inside
You'd see a core of very hard rubber
Surrounded by string though from
your eyes it hides.

The string is wound tightly and seems very solid
But if it was hit hard it would change its shape.
The cover outside is a thin piece of horsehide
That keeps it all in so the ball will not break.

2.

The earth's cover is very much like that horsehide.
It's called the crust. (Not like bread but of rock).
It's between ten miles and thirty miles thick.
That's not very deep though it sounds like a lot.

Under the crust is the part called the mantle?
It's solid, but like the string inside the ball,
It can change shape if it's put under pressure.
It's one thousand eight hundred
miles in all.

The part in the middle is made out of metal.
There are two parts called, together, the core.
The outer core's liquid while the inner one's solid
And the temperature's eight thousand
degrees, maybe more!

The next time you <u>next</u> time you <u>walk</u> somewhere
<u>try</u> to <u>remember</u>
<u>That</u> you are <u>walking</u> on <u>quite</u> a lot <u>more</u>
Than <u>dirt</u>, grass, sand, <u>rocks</u>, and the
<u>critters</u> that <u>live</u> there.
You're <u>walking</u> on the <u>Earth's</u> crust, its
<u>mantle</u> and <u>core</u>.

CHAPTER TWO
What is magma?

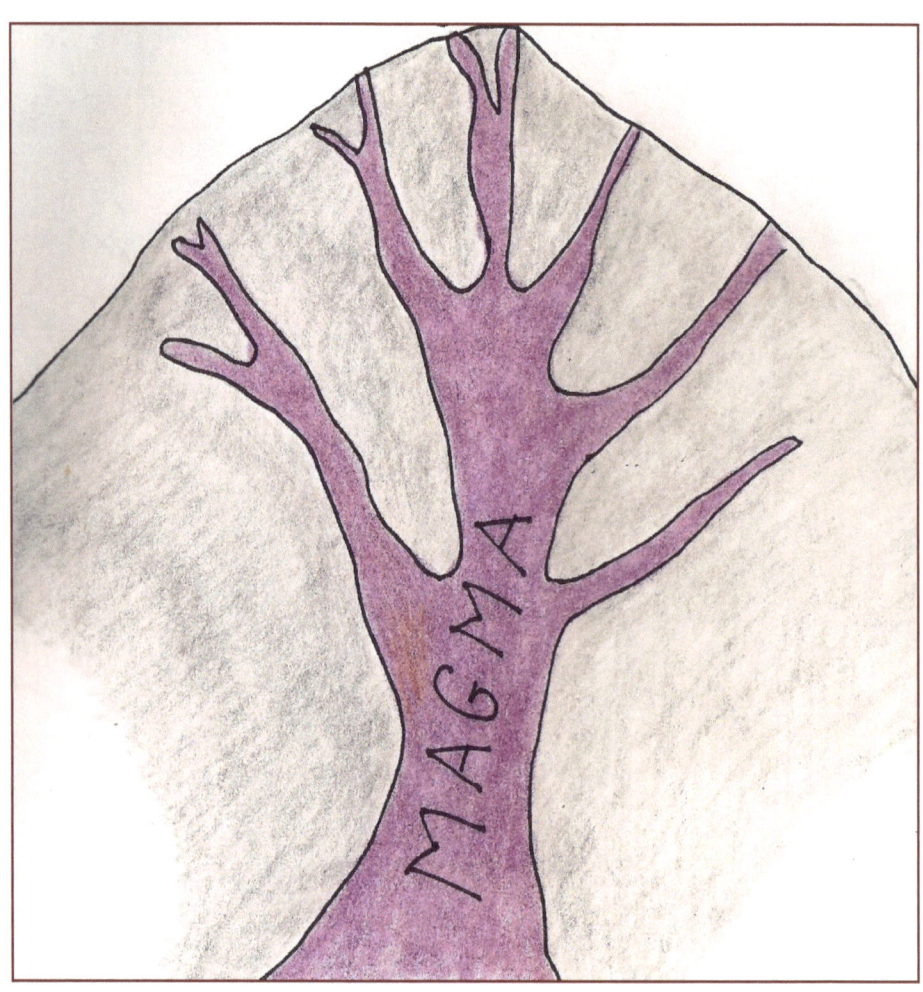

Magma is rock that is so very hot
That it turns to liquid and tries to get out.
If there's a weak spot or crack in the crust
It tries to squeeze through it to make its own route.

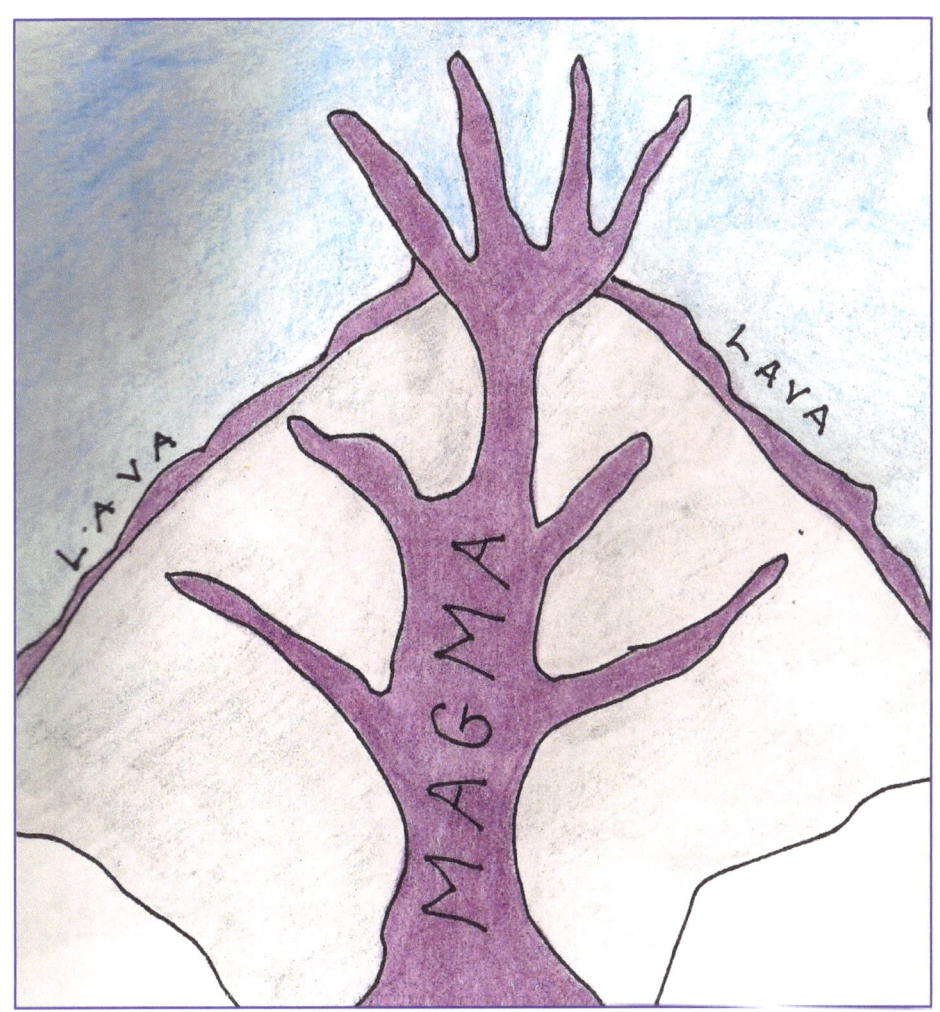

When it <u>reaches</u> the <u>top</u> then there <u>is</u> an ex<u>plosion</u>
That <u>causes</u> a <u>very</u> loud <u>volcanic</u> <u>roar</u>.
Once <u>outside</u> the <u>mountain</u> its <u>new</u> name is <u>lava</u>
<u>As</u> down the <u>mountain</u> this <u>thick</u> lava <u>pours</u>.

6.

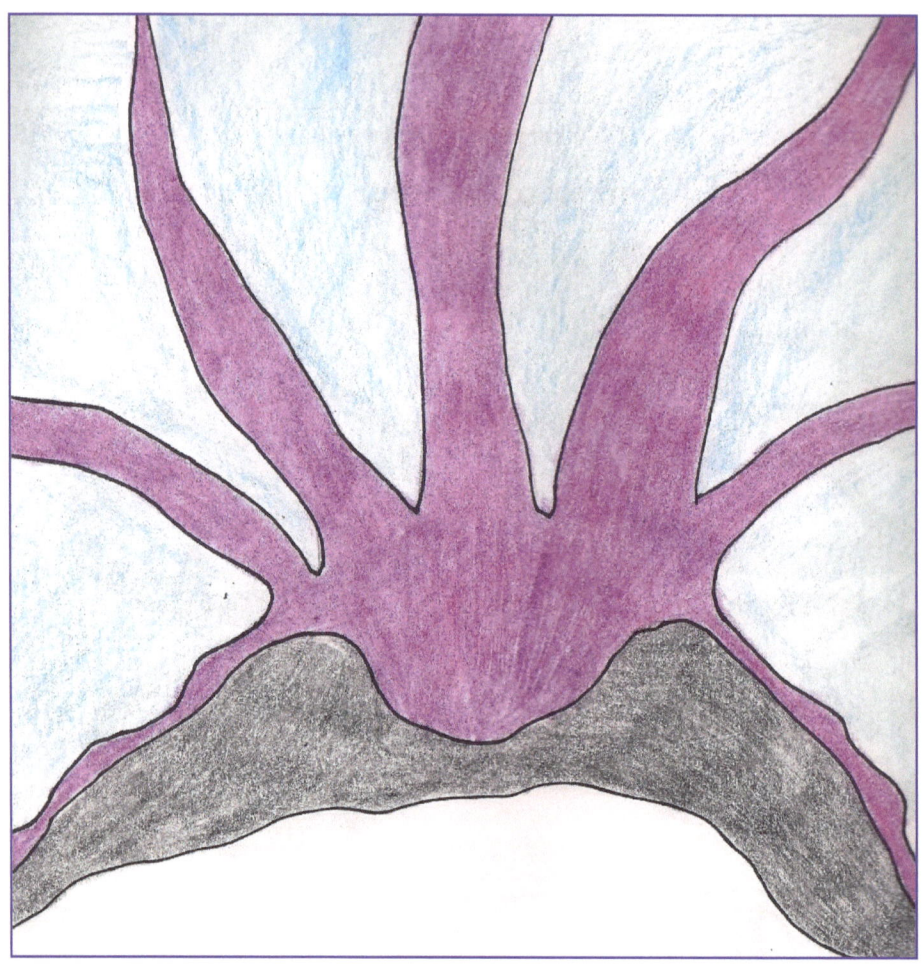

The lava plugs up all the cracks in the mountain
By forming these extremely hot burning pools.
And after the pressure is finally released
The lava will slowly, with time, start to cool.

Whenever the pressure inside of the mountain
Builds up enough to again blow its top,
Once again there'll be a blasting explosion
And out from the mountain comes
more red-hot glop.

7.

CHAPTER THREE
Where does natural gas come from?

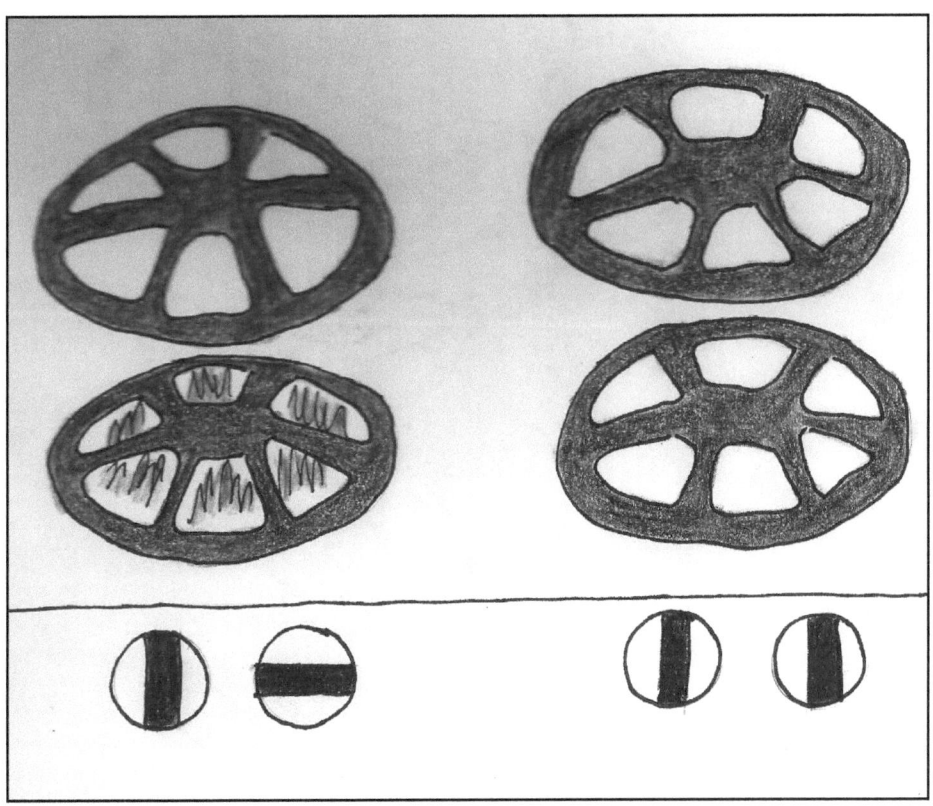

If <u>you</u> have a <u>gas</u> stove <u>inside</u> of your <u>kitchen</u>
And <u>not</u> the <u>electric</u> kinds <u>some</u> people do,
<u>Look</u> at it <u>and</u> see if <u>when</u> it's turned <u>on</u>
The <u>flame</u> appears <u>to</u> be a <u>little</u> bit <u>blue</u>.

8.

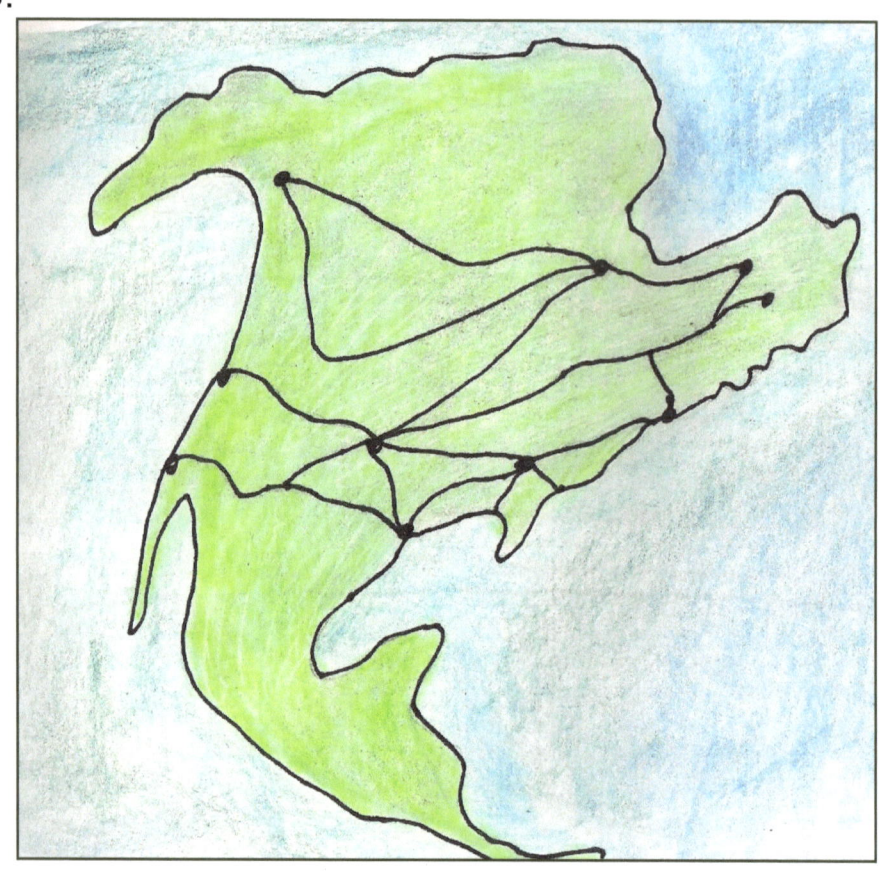

The gas that's inside that stove just
may come from
As far as some hundreds of miles away
Where people drill holes deep inside of the earth
And where all the natural gas pockets lay.

Huge pipes are buried from these natural pockets
For hundreds and hundreds and
hundreds of miles.
All the way through our great nation gas travels
So folks just like us can do cooking in style.

The gas is then stored in both cities and towns
But the journey it makes is still not complete
For the gas in much smaller pipes has to be taken
To all of the houses on your very street.

From there a pipe will go into your home.
It connects to your stove and goes
under the pot
On the stove that your mom uses
to cook the food in.
Natural gas surely does move quite a lot!

10.

CHAPTER FOUR
How is gas made from coal?

Did <u>you</u> know that <u>gas</u> can be <u>made</u> out of <u>coal</u>?
<u>First</u> it is <u>dug</u> from the <u>earth</u> and then <u>sold</u>.

The coal is then dropped down into a long chute
And then into a tank that is basically closed
To make sure it keeps all of the oxygen out,
And the coal, to the air, is never exposed.

If there's no air you cannot have a fire
And so that coal, you see, will never burn.
Keeping the coal from burning all up
Is the coal keeper's biggest concern.

12.

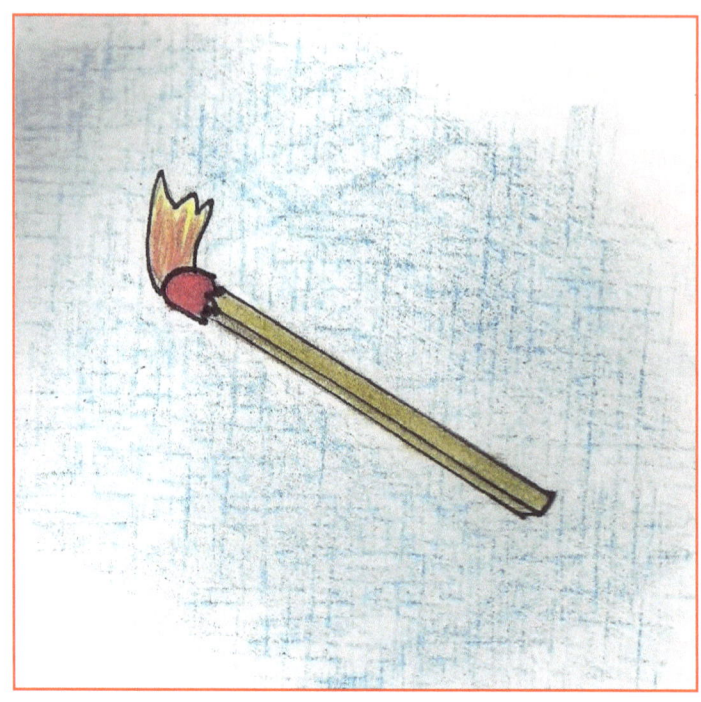

The coal is then heated way up until it
Breaks up into all these small bits and pieces.
Some of the coal turns itself into vapor.
They want the vapor that the coal releases.

The vapor's cooled down,
cleaned up and then sent
By pipes up into your very own home
Or you can just purchase a large storage tank
Or you can get one that the company loans.

Whether your gas is stored in a large tank
Or you get it sent to you via a pipe,
It travels a long way to get to your kitchen
Just so that stove that you have you can light.

13.
CHAPTER FIVE
Why do pots on a gas stove sometimes get black on the bottom?

When you turn on the gas stove in your kitchen
The gas mixes up with all of the air
That passes over the pilot light flame
And causes the gas to suddenly flare.

Gas is a good and dependable fuel
Because it has heat that is stable and steady.
Whenever you see that the flame is bright blue
You know that it's clean and to use it, it's ready.

14.

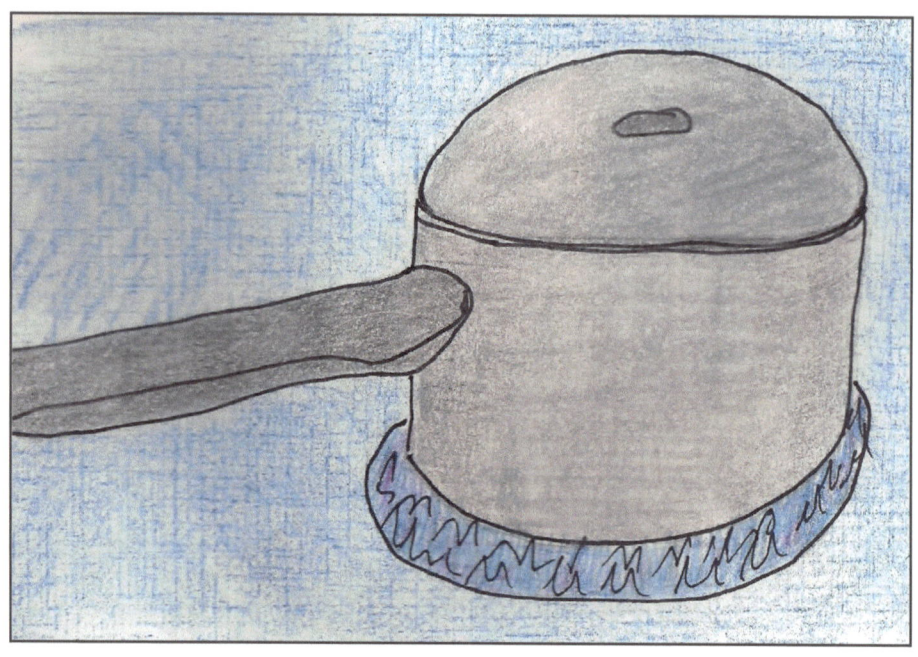

But if the flame is not blue but is yellow
You will find black stuff under your pot.
Because all the coal dust never burned up,
soot on your pot will be what you've got.

Whenever the flame is clearly a blue
You know that the heat that you have will be clean
And your pots and pans you will like
so much better
With bottoms with no soot but glisten with sheen.

CHAPTER SIX
What is a desert like?

In a desert, almost no plant life you'll find.
In a desert you'll find sandy ground.
Most animals hide through the heat of the day.
In a desert you hardly will hear any sounds…

Except, of course, wind as it rushes about
And lifts up the sand on its way,
Whipping and throwing the sand on the rocks
To sculpt them into some very strange shapes.

16.

By <u>day</u> the bright <u>sun</u> is so <u>hot</u> it is <u>fiery.</u>
It <u>beats</u> down up<u>on</u> the <u>desert</u> all <u>day</u>.
<u>It</u> is so <u>hot</u> that <u>people</u> can't <u>live</u> there
For <u>long</u> without <u>trying</u> to <u>lo</u>cate some <u>shade</u>.

But <u>once</u> the sun <u>sinks</u> and it <u>sets</u> in the <u>west</u>
The <u>temper</u>ature <u>drops</u> and you <u>sure</u>ly can <u>freeze.</u>
<u>Any</u>one <u>left</u> outside <u>af</u>ter it's <u>dark</u>
Might <u>shiver</u> and <u>wheeze</u> and s<u>n</u>eeze.

<u>It</u> rarely <u>rains</u> in a <u>desert</u> at <u>all</u>
And <u>that</u> makes the <u>desert</u> seem <u>very</u> much <u>hotter</u>.
So the <u>only</u> plants <u>that</u> you will <u>find</u> in the <u>desert</u>
Are <u>those</u> that can <u>live</u> with<u>out</u> too much <u>water</u>.

18.

But on rare occasions, just once in a while,
A cloudburst will come sending with it a flood.
The cactus are happy and store up the water
And very soon all their flowers will bud.

I never would choose to live in a desert!
That's if it was up to me.
But there are folks that say that they think it's the
Very best place in the whole world to be.

CHAPTER SEVEN
Why does a geyser throw out steam?

A geyser is just like a little volcano
Because it is caused by its inside heat.
There are cracks in the earth and after a rain
Down into those cracks the water will seep.

The water that trickles down inside the earth
Gets so very hot that it turns into steam.
Steam and hot water shoot out of the cracks
In a pretty display that I'd call supreme.

When the steam and the water have
all spurted out
Then the pretty display will suddenly stop
Until more water seeps into the cracks
And water again toward the sky will be shot.

CHAPTER EIGHT
Why is there sand on a beach?

Why do you always find sand on a beach
Where the oceans meet up with the land?
Where does it come from and how did it get there?
I want to understand.

The ocean's the culprit believe it or not
For its waves keep on coming and
don't seem to end.
The pebbles and rocks, the stones
and the seashells
Get pounded by waves again and again.

22.

<u>Some</u> of the <u>waves</u> will <u>carry</u> them <u>off</u>
<u>Only</u> to <u>bring</u> them right <u>back</u> once <u>again</u>.
But <u>always</u> much <u>smaller</u> they <u>tend</u> to be<u>come</u>
<u>Until</u> they are <u>just</u> tiny <u>grains</u> of fine <u>sand</u>.

<u>It</u> takes a <u>lot</u> of <u>bumping</u> and <u>grinding</u>
<u>To</u> make a <u>beach</u>, and <u>that</u> is a <u>fact</u>.
And <u>all</u> that hard <u>work</u> is <u>done</u> just for <u>us</u>
So <u>we</u> can build <u>sand</u> castles <u>or</u> just <u>relax</u>.

CHAPTER NINE
Why is the water in the ocean salty?

There is salt in the dirt. There is salt in the sand.
There is salt in the pebbles and rocks.
So if you play in the dirt you might get
Salt on your feet or your socks.

The river will pick up these small grains of sand
As it makes its way to the sea.
All of the salt in the ocean is dumped.
That process keeps going, you see?

24.

When it <u>rains</u> all the <u>water</u> that's <u>in</u>
lakes and in <u>streams</u>
Get <u>diluted</u> for the <u>raindrops</u> are <u>caught</u>.
<u>Therefore</u>, the <u>water</u> that <u>we</u> drink is <u>good</u>
Be<u>cause</u> it gets <u>rid</u> of the <u>salt</u>.

But the <u>water</u> in <u>oceans</u> have <u>no</u> place to <u>go</u>
So the <u>salt</u> will just <u>keep</u> on collecting.
The <u>rain's</u> not e<u>nough</u> to <u>dilute</u> all that <u>water</u>
And <u>so</u> there's no <u>way</u> for cor<u>recting</u>.

CHAPTER TEN
Why is there always a breeze at the ocean?

All through the day soil and rocks heat up quickly,
While the warming of water is slow.
The dirt on the ground heats just on the surface
While the water gets heated below.

As heated air rises the cooled ocean air
Rushes in to take up the space.
And so all day long the wind comes
from the ocean
And it tends to cool off the place.

26.

But <u>all</u> through the <u>night</u> the earth <u>loses</u> its <u>heat</u>
But the <u>heat</u> in the <u>water</u> <u>remains.</u>
The <u>warm</u> air <u>over</u> the <u>ocean</u> goes <u>up.</u>
The <u>cold</u> air, that <u>space</u> comes to <u>claim</u>.

So the <u>wind</u> blows <u>inland</u> <u>all</u> through the <u>day</u>
But at <u>night</u> it <u>turns</u> and <u>heads</u> to the <u>sea.</u>
That's <u>why</u> there's <u>always</u> a <u>breeze</u> at the ocean
For the <u>air</u> is <u>coming</u> or <u>going</u>, you <u>see</u>?

CHAPTER ELEVEN
What holds an island up?

When you look at an island it seems to be floating
Along on the gentle waves.
But when you read this and find out the truth
You simply will be amazed.

28.

An island, you see, is actually a mountain.
Only the top's what you see.
The rest of the island is under the water,
Down under the ocean or sea.

Sometimes small islands can seem to just vanish
Because they are nowhere in sight.
But they're still there, just under the water,
If the water rises too high.

CHAPTER TWELVE
How were caves formed?

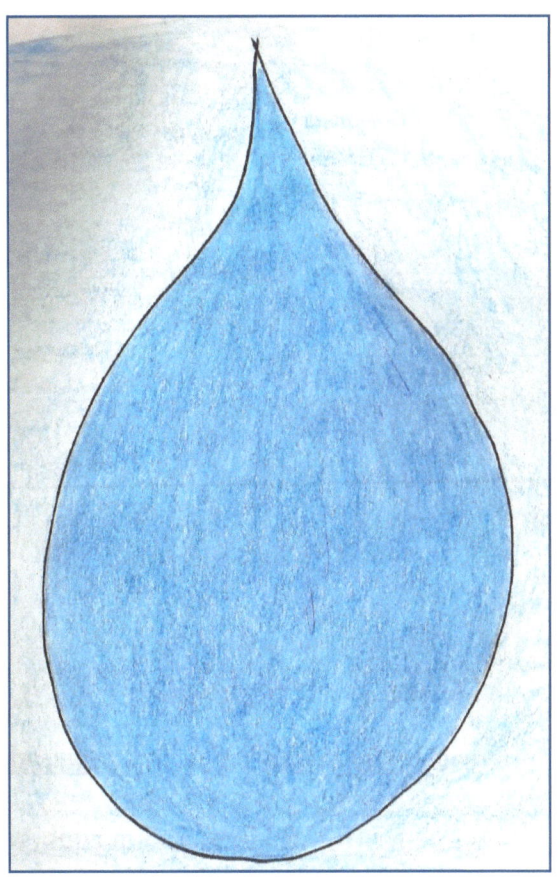

Have you ever
gone on a
spelunking trip
Way down inside
of a cave?...
With icicles made
out of stone
hanging down
Or rising up from
stony graves?....

Where delicate
colored stone
petals will grow
Between the rough
rocks and the
cracks?.....
And where rivers flow and waterfalls tumble
Down into a world that is truly pitch black?

A cave gets its start because of the rain
That seeps way down into the ground,
Slowly dissolving rocks to bits of stone.
That's where today a cave can be found.

30.

For thousands of years the billions of drips
Going inside a stony cave
Is like a magician's magical wand
Forming a fairyland's wondrous arcade.

Stalagmites will rise from the floor
formed by dripping,
And dripping and dripping year after year.
Stalactites that form will hang
down from the ceiling
Like beautiful chandeliers.

And so over millions and millions of years
(Because they build slowly, that's certain)
The mites and the tites will finally touch,
Forming a beautiful sculptured, stone curtain.

CHAPTER THIRTEEN
How can wind and rain carve rocks?

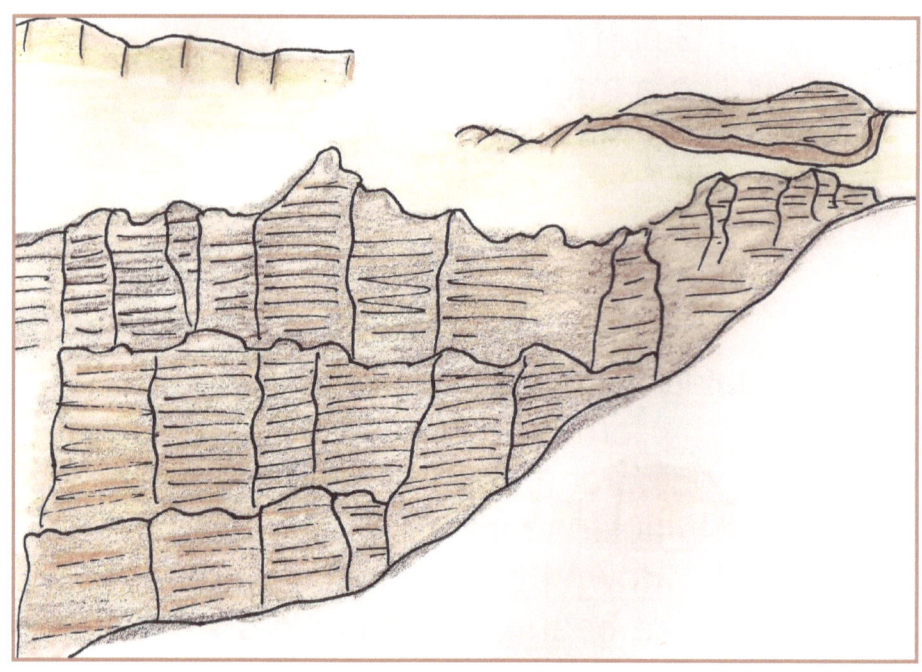

The w<u>ind </u>and the w<u>ater</u> work
<u>like</u> partnered s<u>culpt</u>ors.
<u>Constantly they</u> are at <u>work</u>.
The <u>beauty</u> that <u>we</u> can see <u>in</u> the Grand <u>Canyon</u>
Is <u>surely</u> not <u>simply</u> a <u>quirk</u>.

It <u>took </u>many <u>millions of years </u>for the <u>rain</u>
to <u>pour</u> and <u>carry</u> a<u>way</u>
The <u>tiniest bits</u> of the <u>limestone rock</u>
That e<u>ventually made </u>that dis<u>play</u>.

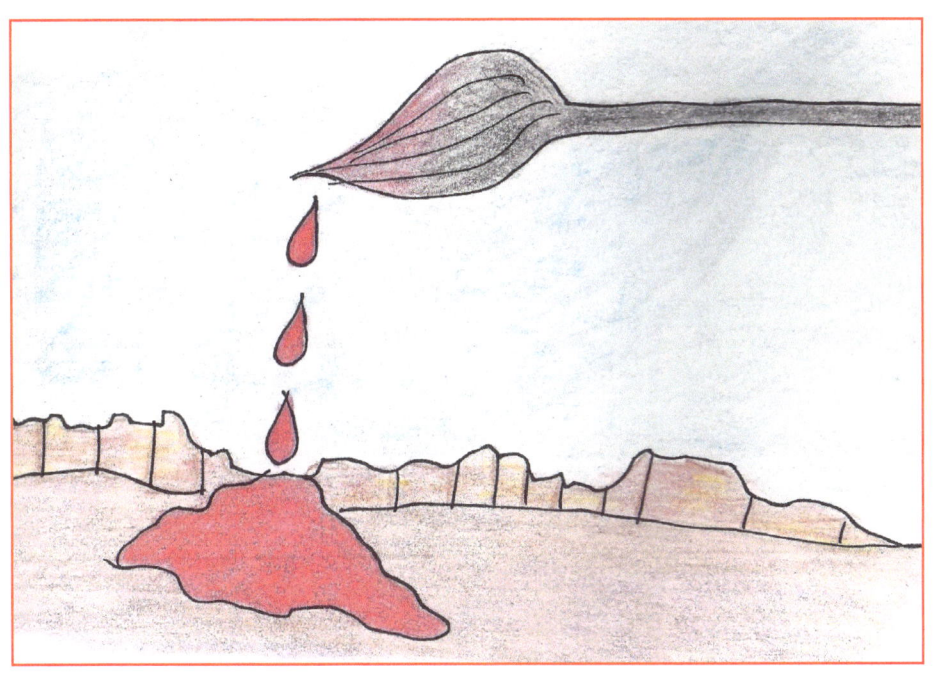

Look at a picture of the Painted Desert
With its shapes so weird and yet grand.
That desert was formed by the countless, tiniest,
windblown grains of sand
That eventually carved out a wonderful fairyland
for all of us to see,
Then painted it all in soft pastel colors.
It's surely amazing to me!

CHAPTER FOURTEEN
What are wind caves?

On a rocky cliff or the side of a hill
Where the winds blow steady and are never still;
Where a layer of shale that is soft is wedged
Between the hard rock of a sandstone bed:

This is where wind caves eventually form;
Where the strong wind is whipped from an ocean's storm.
Currents of air swirl over the face
Of the hillside, scooping shale out from the base.

Further and further, and deeper it goes,
Shaping a cave as the strong wind blows.
For thousands of years the wind whistled and groaned
Making cavemen happy while building their homes.

CHAPTER FIFTEEN
What was the ice age?

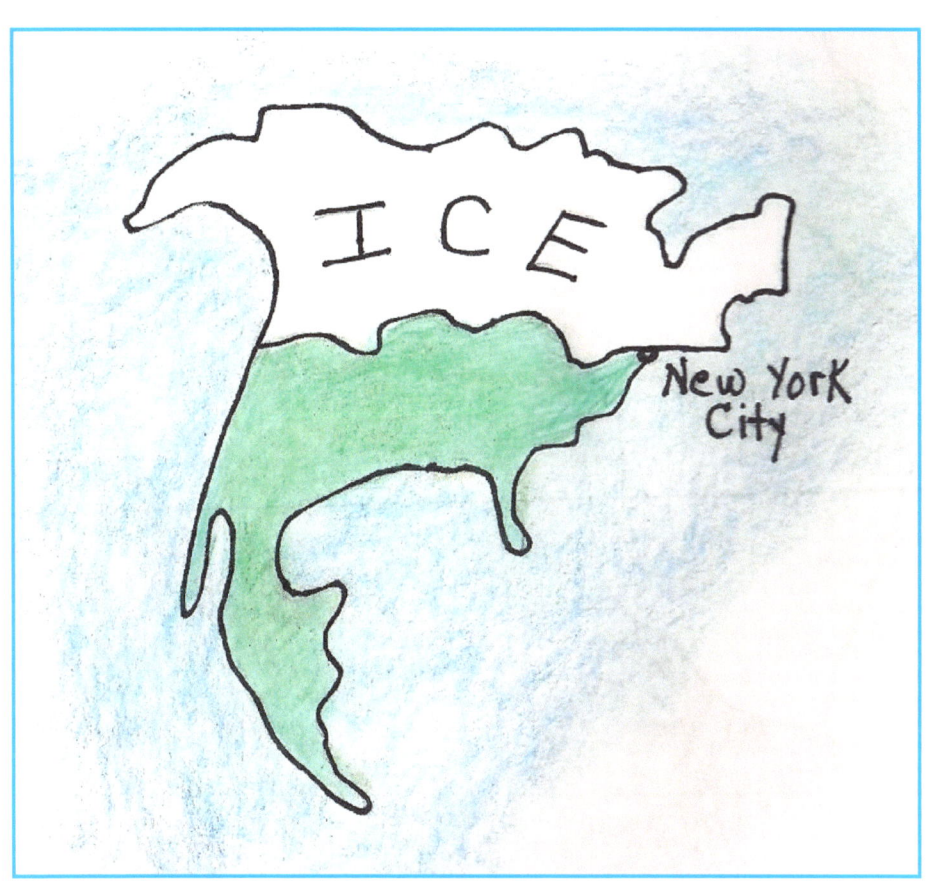

For <u>thous</u>ands of <u>years</u> the earth <u>was</u> really <u>cold</u>…
A <u>fourth</u> of the <u>earth</u> buried <u>in</u> ice and <u>snow,</u>
From <u>ice</u>-covered <u>Canada</u> <u>and</u> the Great <u>Lakes,</u>
<u>Down</u> through the <u>western</u> plains
<u>and</u> northern <u>states</u>.

I am so glad that the ice age has left us
Because it's Alaska that's where
We like to think live the critters like seals
and the fluffiest, white polar bears.

CHAPTER SIXTEEN
What is a glacier?

A glacier's a very large river of ice
That flows down a mountainside
Cutting itself a path that is made
From the rocks and the soil that it rides.

Ice starts to form on the top of a mountain,
Where it snows year after year,
Compressing the snow down into solid ice;
A river of ice thick and sheer.

When the glacier gets too heavy a
chunk breaks away
Becoming an iceberg.
Quite heavy it weighs!

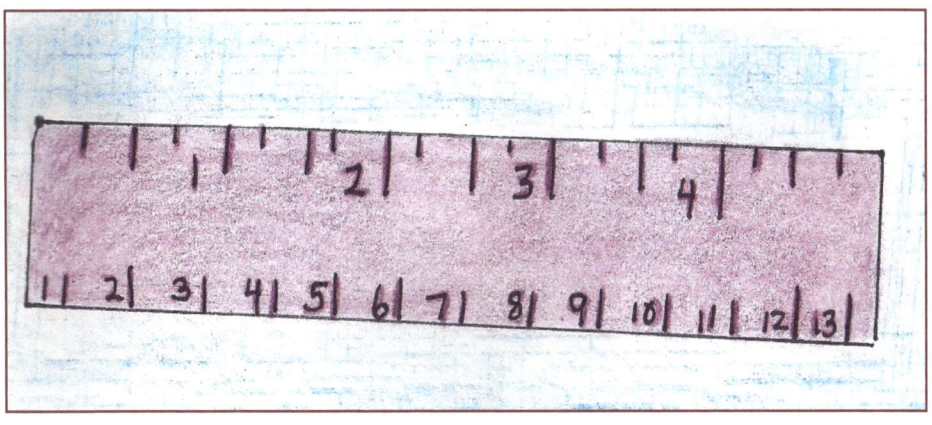

A glacier moves slowly, just inches each day
Until warmer weather towards it makes its way.
When it starts to melt the rivers are filled
Because glacier water into oceans spill.

Because global warming makes everything warm
The melting of glaciers is causing alarm!
The glaciers are melting at such a fast pace
That if something's not done they will
melt clear away!

CHAPTER SEVENTEEN
What are ice caves?

Some of you may have been inside a cave
But don't you think it would be great
To go inside a cave that was made out of ice?
I wouldn't hesitate!

In the European mountains ages ago
The climate was warmer than right now, and so
Underground rivers through rock tunneled through
Making a cave, though it took long to do.

But the Ice Age turned water into ice instead.
And today they exist as iced riverbeds.
The water that froze formed into frozen lakes
As smooth as a rink where you might go ice-skate.
I'd surely love seeing this with my own eyes…
A crystal, jeweled, wonderland made out of ice!

CHAPTER EIGHTEEN
Why does a compass point north?

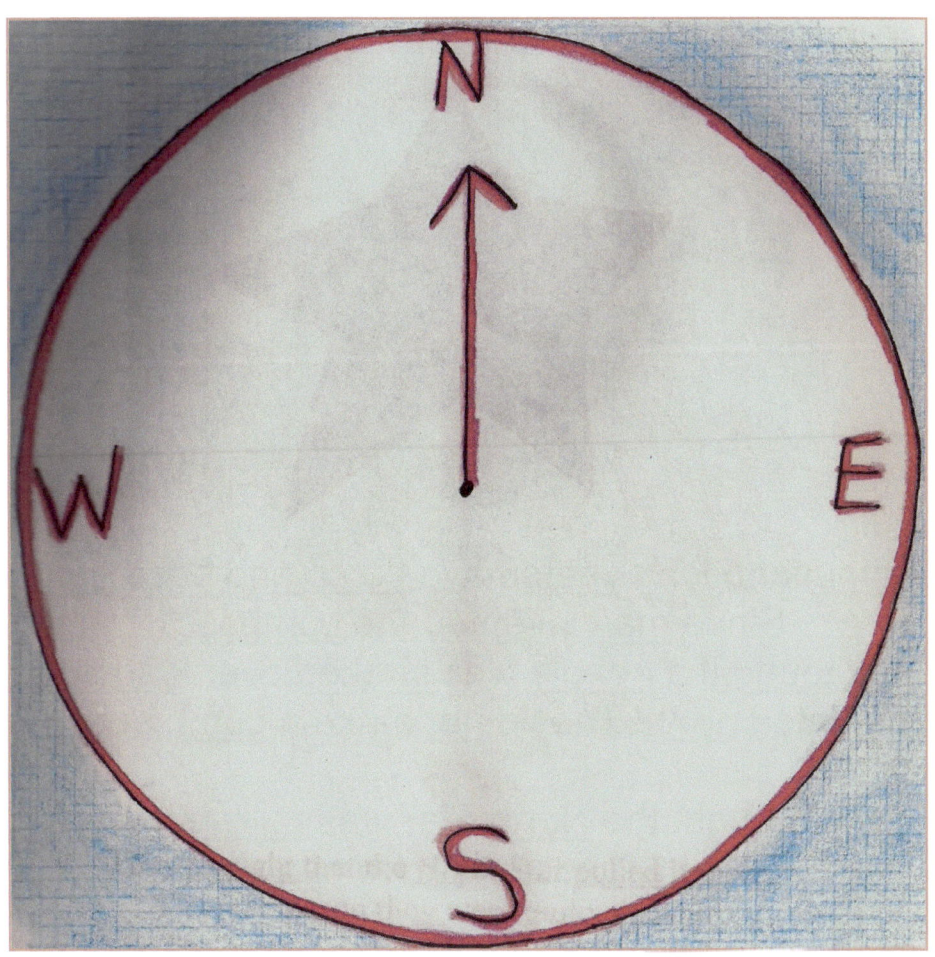

If you're ever alone in the woods and get lost
And your way back home you find's concealed.
You'd know which way was north if
you had a compass
That's made from magnetized steel.

42.

A <u>thou</u>sand long <u>years</u> have <u>come</u> and have <u>gone</u>
Since <u>man</u> in<u>ven</u>ted the <u>compass</u>.
They <u>found</u> if a <u>needle</u> was <u>stroked</u> over <u>loadstone</u>
<u>It</u> pointed <u>north</u>! They'd <u>not</u> make a <u>dumb</u> guess.

<u>They</u> thought the <u>North</u> Star was <u>pulling</u> it <u>there</u>.
The <u>real</u> reason <u>now</u> we can <u>share</u>.
The <u>earth</u> is a <u>magnet</u> and <u>it</u> has <u>control</u>.
The <u>needle</u> points <u>to</u> the mag<u>netic</u> North <u>Pole</u>.

So, <u>if</u> you are <u>lost</u>, but a <u>compass</u> you <u>own</u>,
It's <u>almost</u> as <u>good</u> as <u>having</u> a <u>phone</u>.
It <u>shows</u> you where <u>north</u> is so <u>then</u> you can <u>find</u>
<u>East</u>, south and <u>west</u> and be <u>home</u> right on <u>time</u>.

LOOK FOR THESE OTHER BOOKS FROM THE
SCIENCE WITH A BEAT
SERIES

OUT BEYOND

WEATHER

WATER WORLD

ELECTRICITY

www.ingramcontent.com/pod-product-compliance
Lightning Source LLC
Chambersburg PA
CBHW040927180526

45159CB00002BA/643